SPARKS OF LIFE

Chemical Elements that Make Life Possible

CALCIUM

by

Jean F. Blashfield

Austin, Texas

Special thanks to our technical consultant,
Jeanne Hamers, Ph.D.,
formerly with the Institute of Chemical Education,
Madison, Wisconsin

Development: Books Two, Delavan, Wisconsin
Graphics: Krueger Graphics, Janesville, Wisconsin
Interior Design: Peg Esposito
Photo Research and Indexing: Margie Benson

Raintree Steck-Vaughn Publisher's Staff:
Publishing Director: Walter Kossmann
Design Manager: Joyce Spicer
Project Editor: Frank Tarsitano
Electronic Production: Scott Melcer

Library of Congress Cataloging-in-Publication Data:
Blashfield, Jean F.
Calcium / by Jean F. Blashfield.
p. cm. — (Sparks of life)
Includes bibliographical references (p. -) and index.
Summary: Presents the basic concepts of this metallic element which is found in the stone used to make buildings and streets and which is an important mineral needed by the body.
ISBN 0-8172-5040-9
1. Calcium — Juvenile literature. [1. Calcium.] I. Title. II. Series: Blashfield, Jean F. Sparks of life.
QD181.C2B58 1999 98-4518
546' .393—dc21 CIP
AC

Printed and bound in the United States
1 2 3 4 5 6 7 8 9 LB 02 01 00 99 98

PHOTO CREDITS: American Electric Company 49; ARS–Information Staff cover; Bethlehem Steel 51; ©Biophoto Associates/Science Source 36; Corbis 11; Courtesy of Church & Dwight Co., Inc. 57; Exxon Company, USA 42; Georgia Marble Company 20; ©Twice Gibson, Pacific Stock cover; ©1995 Hossler, Ph.D./Custom Medical Stock Photo 33; ©Bruce Iverson 18, 46; JLM Visuals 15, 16, 25, 39, 54; ©Greg Johnston, International Stock 44; Barbara Krause 22; ©Prof. P. Motta/Dept. of Anatomy/ University "La Sapienza," Rome/Science Photo Library cover; Mark Newman, International Stock 24; ©Science Photo Library 13; Scalewatcher 40; ©Secchi-Lecaque/Roussel-UCLAF/CNRI/Science Photo Library 30; ©Charles D. Winters/Photo Researchers 14; U.S. Capitol, Office of the Curator 47; ©Ted Wood 23; Wisconsin Milk Marketing Board 35.

CONTENTS

Periodic Table of the Elements

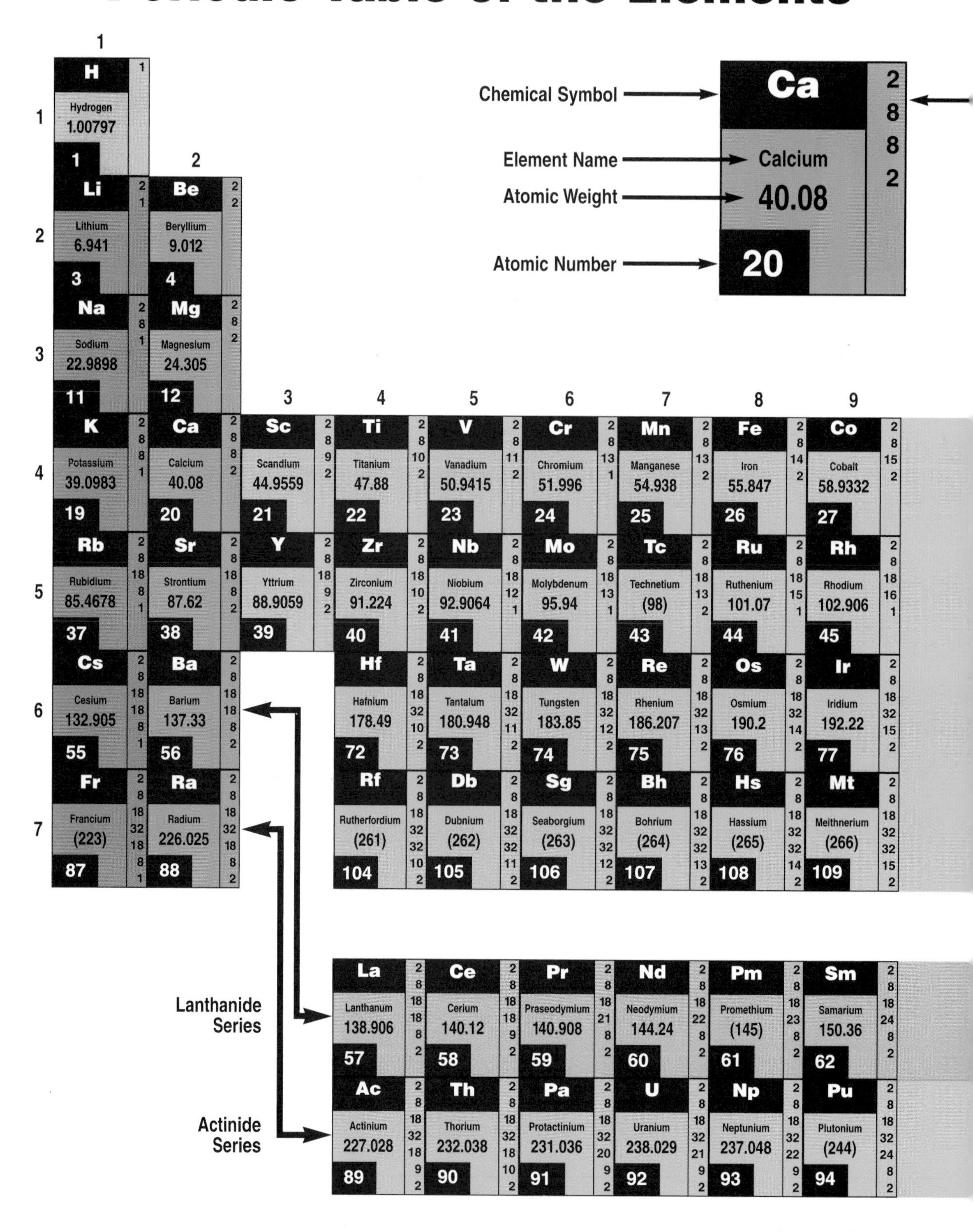

Number of electrons in each shell,
beginning with the K shell, top.

See next page for explanations.

10	11	12	13	14	15	16	17	18
								He Helium 4.0026 2 2
			B Boron 10.81 5 2 3	C Carbon 12.011 6 2 4	N Nitrogen 14.0067 7 2 5	O Oxygen 15.9994 8 2 6	F Fluorine 18.9984 9 2 7	Ne Neon 20.179 10 2 8
			Al Aluminum 26.9815 13 2 8 3	Si Silicon 28.0855 14 2 8 4	P Phosphorus 30.9738 15 2 8 5	S Sulfur 32.06 16 2 8 6	Cl Chlorine 35.453 17 2 8 7	Ar Argon 39.948 18 2 8 8
Ni Nickel 58.69 28 2 8 16 2	Cu Copper 63.546 29 2 8 18 1	Zn Zinc 65.39 30 2 8 18 2	Ga Gallium 69.72 31 2 8 18 3	Ge Germanium 72.59 32 2 8 18 4	As Arsenic 74.9216 33 2 8 18 5	Se Selenium 78.96 34 2 8 18 6	Br Bromine 79.904 35 2 8 18 7	Kr Krypton 83.80 36 2 8 18 8
Pd Palladium 106.42 46 2 8 18 18	Ag Silver 107.868 47 2 8 18 18 1	Cd Cadmium 112.41 48 2 8 18 18 2	In Indium 114.82 49 2 8 18 18 3	Sn Tin 118.71 50 2 8 18 18 4	Sb Antimony 121.75 51 2 8 18 18 5	Te Tellurium 127.6 52 2 8 18 18 6	I Iodine 126.905 53 2 8 18 18 7	Xe Xenon 131.29 54 2 8 18 18 8
Pt Platinum 195.08 78 2 8 18 32 17 1	Au Gold 196.967 79 2 8 18 32 18 1	Hg Mercury 200.59 80 2 8 18 32 18 2	Tl Thallium 204.383 81 2 8 18 32 18 3	Pb Lead 207.2 82 2 8 18 32 18 4	Bi Bismuth 208.98 83 2 8 18 32 18 5	Po Polonium (209) 84 2 8 18 32 18 6	At Astatine (210) 85 2 8 18 32 18 7	Rn Radon (222) 86 2 8 18 32 18 8
(Uun) (Ununnilium) (269) 110 2 8 18 32 32 17 1	(Unu) (Unununium) (272) 111 2 8 18 32 32 18 1	(Uub) (Ununbium) (277) 112 2 8 18 32 32 18 2						

Alkali Metals | Transition Metals | Nonmetals | Metalloids | Lanthanide Series

Alkaline Earth Metals | Other Metals | Noble Gases | Actinide Series

COLOR KEYS

Eu Europium 151.96 63 2 8 18 25 8 2	Gd Gadolinium 157.25 64 2 8 18 25 9 2	Tb Terbium 158.925 65 2 8 18 27 8 2	Dy Dysprosium 162.50 66 2 8 18 28 8 2	Ho Holmium 164.93 67 2 8 18 29 8 2	Er Erbium 167.26 68 2 8 18 30 8 2	Tm Thulium 168.934 69 2 8 18 31 8 2	Yb Ytterbium 173.04 70 2 8 18 32 8 2	Lu Lutetium 174.967 71 2 8 18 32 9 2
Am Americium (243) 95 2 8 18 32 25 8 2	Cm Curium (247) 96 2 8 18 32 25 9 2	Bk Berkelium (247) 97 2 8 18 32 26 9 2	Cf Californium (251) 98 2 8 18 32 28 8 2	Es Einsteinium (254) 99 2 8 18 32 29 8 2	Fm Fermium (257) 100 2 8 18 32 30 8 2	Md Mendelevium (258) 101 2 8 18 32 31 8 2	No Nobelium (259) 102 2 8 18 32 32 8 2	Lr Lawrencium (260) 103 2 8 18 32 32 9 2

A Guide to the Periodic Table

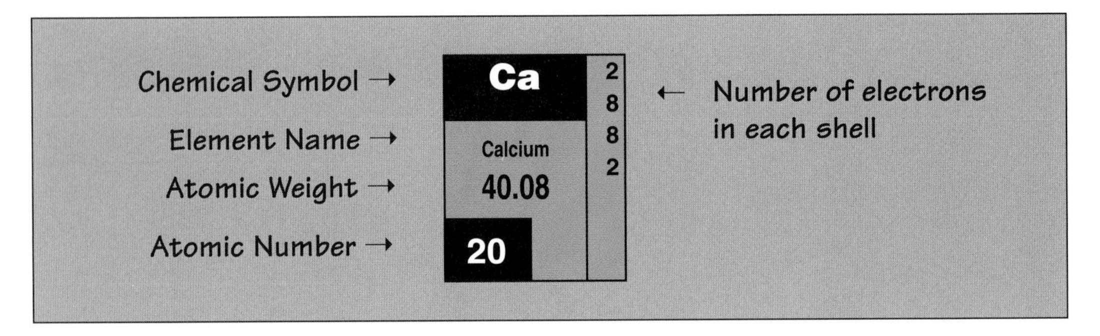

Symbol = an abbreviation of an element name, agreed on by members of the International Union of Pure and Applied Chemistry. The idea to use symbols was started by a Swedish chemist, Jöns Jakob Berzelius, about 1814. Note that the elements with numbers 110, 111, and 112, which were "discovered" in 1996, have not yet been given official names.

Atomic number = the number of protons (particles with a positive electrical charge) in the nucleus of an atom of an element; also equal to the number of electrons (particles with a negative electrical charge) found in the shells, or rings, of an atom that does not have an electrical charge.

Atomic weight = the weight of an element compared to carbon. When the Periodic Table was first developed, hydrogen was used as the standard. It was given an atomic weight of 1, but that created some difficulties, and in 1962, the standard was changed to carbon-12, which is the most common form of the element carbon, with an atomic weight of 12.

The Periodic Table on pages 4 and 5 shows the atomic weight of carbon as 12.011 because an atomic weight is an average of the weights, or masses, of all the different naturally occurring forms of an atom. Each form, called an isotope, has a different number of neutrons (uncharged particles) in the nucleus. Most elements have several isotopes, but chemists assume that any two samples of an element are made up of the same mixture of isotopes and thus have the same mass, or weight.

Electron shells = regions surrounding the nucleus of an atom in which the electrons move. Historically, electron shells have been described as orbits similar to a planet's orbit. But actually they are whole areas of a specific energy level, in which certain electrons vibrate and move around. The shell closest to the nucleus, the K shell, can contain only 2 electrons. The K shell has the lowest energy level, and it is very hard to break its electrons away. The second shell, L, can contain only 8 electrons. Others may contain up to 32 electrons. The outer shell, in which chemical reactions occur, is called the valence shell.

Periods = horizontal rows of elements in the Periodic Table. A period contains all the elements with the same number of orbital shells of electrons. Note that the actinide and lanthanide (or rare earth) elements shown in rows below the main table really belong within the table, but it is not regarded as practical to print such a wide table as would be required.

Groups = vertical columns of elements in the Periodic Table; also called families. A group contains all elements that naturally have the same number of electrons in the outermost shell or orbital of the atom. Elements in a group tend to behave in similar ways.

Group 1 = alkali metals: very reactive and so never found in nature in their pure form. Bright, soft metals, they have one valence electron and, like all metals, conduct both electricity and heat.

Group 2 = alkaline earth metals: also very reactive and thus don't occur pure in nature. Harder and denser than alkali metals, they have two valence electrons that easily combine with other chemicals.

Groups 3–12 = transition metals: the great mass of metals, with a variable number of electrons; can exist in pure form.

Groups 13–17 = transition metals, metalloids, and nonmetals. Metalloids possess some characteristics of metals and some of nonmetals. Unlike metals and metalloids, nonmetals do not conduct electricity

Group 18 = noble, or rare, gases: in general, these nonmetallic gaseous elements do not react with other elements because their valence shells are full.

THE ALWAYS-CHANGING ELEMENT

Calcium is a metallic element that almost no one ever sees in its pure form, yet it is part of the main ingredient in the stone used to make our buildings and streets. When joined to oxygen (O, element #8), it makes calcium oxide (CaO), which can burn the skin, yet calcium is an important mineral needed by our bodies.

Pure calcium is a fairly soft silver metal and it can conduct electricity. However, pure calcium never occurs in nature. It can be isolated in the laboratory, but it rarely exists for long because it easily reacts with water or air to form calcium compounds.

Calcium is the fifth most plentiful element in the Earth's crust. More than 3 percent of the crust is calcium, mostly in the form of limestone and chalk.

The chemical symbol for calcium is Ca. The element has atomic number 20 and an atomic weight of 40.08. It belongs to Group 2, the second column of elements shown on the Periodic Table.

The Group 2 elements are called alkaline earth metals. Alkali is another name for certain types of chemicals that accept protons from other sources. Chemists of old gave the name "earth" to any material that was usually found in the ground and that did not seem to react to heat or dissolve in water. That name was not accurate for calcium, but the name for the category persisted.

Two Important Electrons

The alkaline earths are metals that have two electrons in their outer, or valence, electron shell. These elements readily give up those two electrons. When that happens, the next inner shell becomes the valence shell. Since that shell has its full number of electrons, the atom becomes stable.

Two alkaline earths are lighter than calcium: beryllium (Be, element #4) and magnesium (Mg, #12); and three are heavier: strontium (Sr, #38), barium (Ba, #56), and radium (Ra, #88).

Calcium is the most important of the substances nutritionists call "minerals" that our bodies need to stay healthy. In fact, it is a macromineral, one that makes up a larger percentage of the human body than some of the other minerals because it is the primary ingredient of bones and teeth.

Because calcium atoms so easily give up their valence electrons to other atoms, calcium atoms are often found as positive ions. Ions are atoms or molecules that are missing or have extra electrons. The positive calcium ion is indicated as Ca^{2+}, meaning that two electrons in the outer shell are gone. The calcium ion is an atom that has two more protons than electrons, thus giving it a positive electrical charge.

The Stone Forest in China was made when softer materials eroded away leaving pinnacles of limestone.

Calcium ions react with anions (negative ions) of other substances in solutions, the new calcium compounds often form solids that precipitate (drop to the bottom) out of the solution and are called precipitates. Precipitation is often used as a means of separating an impurity from a liquid. The impurity reacts with calcium and then settles to the bottom of the container holding the liquid. A precipitate forms when the impurity that is dissolved becomes part of a compound that cannot dissolve.

In Ancient Times

The ancient Greeks and Romans knew that if the common rock we call limestone were heated, it changed into a slightly different mineral. This mineral was a type of lime, a white chalk-like material that the Romans called *calx*. That name led to our element name, calcium. It also led to our adjective calcareous, which is used to describe anything that contains calcium carbonate, $CaCO_3$, the main mineral in limestone.

The Romans' calx was actually the chemical we call calcium oxide. This type of lime is a very useful. The ancient Greeks and Romans used it both to help crops grow and as mortar, a material used to hold layers of bricks or stones together.

As noted, the main mineral making up limestone is calcium carbonate. (Carbon is element #6, with symbol C.) One form of calcium carbonate is called calcite. This form of $CaCO_3$ is crystalline, meaning that the molecules have a regularly repeating internal arrangement of its atoms and often external flat faces. Calcite can occur in many different lustrous colors. The tomb of the Egyptian pharaoh called King Tut contained a beautiful oil lamp carved out of calcite.

The Egyptian name for calcite was alabaster. But calcite is different from the mineral we now call alabaster, which is calcium sulfate ($CaSO_4$) or gypsum. The Egyptians did use gypsum to make the hard bowls in which grains and other minerals were ground up. They also used it as a wall-plastering material. In fact, the inside of King Tut's tomb inside a pyramid is coated with gypsum.

Joseph Black

Early Exploration

Even the ancients knew that when limestone is heated, the lime that remains afterward weighs less than the original limestone. They did not know why.

In 1754, Joseph Black, a British chemist, showed that heated limestone loses weight because an invisible gas is driven off into the air, leaving calcium oxide. He also showed that if the calcium oxide were then exposed to air, the invisible gas apparently reentered the lime, changing it back into limestone.

Black called this invisible gas "fixed air" because it could be "fixed" into a solid. He thus demonstrated that "fixed air" was a normal part of the atmosphere. We now know that Black's "fixed air" was carbon dioxide, CO_2.

Early chemists also believed that lime itself was an element, a substance that could not be broken down further. French chemist Antoine Lavoisier, who worked in the late 1700s, disagreed. He suggested that lime and several similar substances are oxides—substances made up of an element combined with oxygen. He thought it was "probable that we know only part of the metallic substances which exist in Nature."

Isolating Calcium

Lavoisier was just guessing that lime was an oxide. His guess was not proved correct until 1808. In that year, English chemist Humphry Davy was feeling triumphant for having isolated (separated from their compounds) the elements sodium (Na, element #11) and potassium (K, #19). These important alkali metals make up part of Group 1 of the Periodic Table. Davy decided to tackle isolating the alkaline earth elements next.

To isolate alkali metals, Davy had set up an electrolysis experiment. Electrolysis is a method of using electricity flowing through a fluid substance to cause the substance to decompose. He submerged two metal plates, called electrodes, in a fluid. One plate had a positive electrical charge. The other had a negative electrical charge. As a source of electricity, Davy constructed the biggest battery ever made up to that time. When he connected the battery to the plates, the molecules making up the fluid separated, or decomposed, into positive and negative ions. The positively charged particles were drawn to the negative electrode, where they gained electrons. The negatively charged

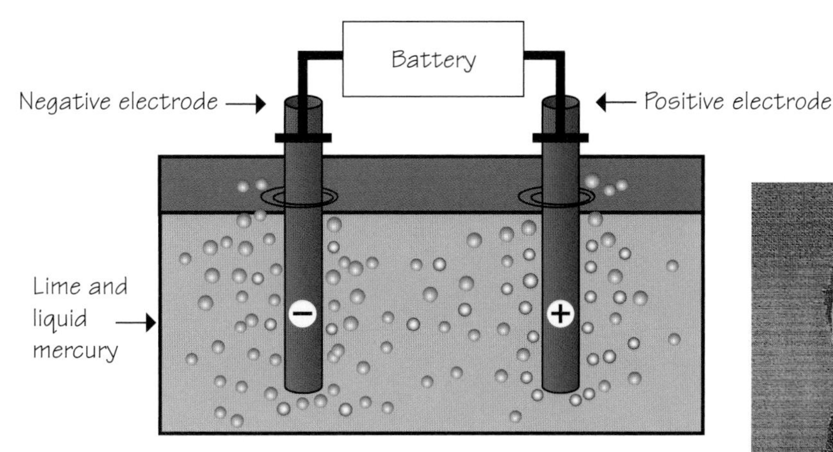

A basic electrolysis apparatus (above) was used by Sir Humphry Davy (right) in 1808 to isolate calcium and other elements.

particles were drawn to the positive electrode, where they lost electrons.

That process worked fine for sodium and potassium compounds in solution, but Davy could not get electrolysis to work with the lime. He tried many different experiments, but none of them worked—or if they did, he couldn't obtain enough calcium to prove he had isolated the element.

Then Davy received a letter from Swedish chemist Jöns Jakob Berzelius. Berzelius described some work he was doing with mercury, which is a fluid. (Mercury is element #80, with symbol Hg for *hydrargyrum*, which means "liquid silver.") Davy decided to mix lime, CaO, with liquid mercury.

Once again using his huge battery to power the electrolysis, Davy succeeded in roughly separating calcium from the other elements. "Roughly" because he failed to make pure calcium. That achievement required another half century of experimentation. Robert Bunsen, the German chemist who invented the Bunsen burner used in chemistry laboratories, succeeded in the 1860s in isolating pure calcium.

Today, electrolysis is still used to collect calcium by decomposing calcium chloride, $CaCl_2$. (Chlorine is Cl, element #17.) With another chemical added to lower its melting point, calcium chloride becomes molten and can serve as the electrolysis fluid, or electrolyte. Positively charged calcium ions are drawn to the negative electrode where they gain electrons and become pure calcium. However, there can never be very much pure calcium collected because it reacts so quickly with any other substance around it. It's fortunate that pure calcium has almost no uses because it reacts so readily with oxygen and water.

A sample of pure calcium metal

ANCIENT SEAS AND MODERN BUILDINGS

One of the most widespread and useful types of rocks in Earth's crust is limestone. It makes up between 9 and 10 percent of the Earth's sedimentary rocks. Sedimentary rocks were formed by solid material settling to the bottom of a lake, river, or ocean. These deposits, called sediment, hardened under heat and pressure caused by layers of other matter settling on top over millions of years.

Limestone can vary from being fine-grained to being very coarse. One variety called chalk can crumble easily. This type of limestone is the main ingredient of the so-called white cliffs of Dover in England. The rocks of these tall cliffs on the English Channel break off in chunks as the sea undermines them.

The chalk cliffs of southern England, near Dover, are a type of limestone made from sediments formed by the remains of millions of shelled animals.

The shells of millions of animals that lived in ancient seas made the limestone we use today. This is a photo of an exhibit of ancient sea life at the Field Museum of Natural History in Chicago.

The chalk of the white cliffs of Dover and similar places is not the same chalk we use to write on chalkboards. That chalk is a manufactured product with the chemical name calcium sulfate ($CaSO_4$).

As we have seen, the main mineral in limestone is calcium carbonate. Calcium carbonate occurs naturally in seawater. It is among the minerals that make seawater different from fresh water.

How the Mineral Got into the Sea

Feldspar is a type of rock that usually contains calcium, silicon (Si, element #14), and aluminum (Al, #13). Feldspars are igneous (fiery) rocks. Igneous rocks were formed inside the earth as molten material and came to or near the surface, perhaps as lava. Once such rocks are exposed to air, they begin to weather, or wear away.

Water from rain and carbon dioxide from the atmosphere can change calcium-rich feldspar into clay and various ions. Two of these ions are positively charged calcium ions (Ca^{2+}) and negatively charged carbonate ions (CO_3^{2-}). Both of these ions exist in great abundance in our planet's waters.

These two ions are attracted to each other because the charge on one is negative and the charge on the other is positive. They react together, forming electrically neutral calcium carbonate ($CaCO_3$):

$$Ca^{2+} + CO_3^{2-} \rightarrow CaCO_3$$

Early in the Earth's history, during the formation of the planet's crust, huge seas evaporated, leaving behind deep layers of calcium carbonate, often mixed with other minerals, and making limestone. Almost unlimited supplies of limestone have made it a major building stone around the world.

Another Source of Limestone

Many of the billions and trillions of sea creatures that inhabited the seas over millions of years used the calcium carbonate in seawater to construct shells. When these ancient creatures died, their shells accumulated in layers that eventually became many feet thick. These shells also contributed to the layers of limestone that may be 1 mile (1.6 km) or more thick in the crust of the Earth.

The ground beneath much of the state of Florida is actually a deposit of calcium carbonate from the shells of such creatures. But the deposits aren't necessarily on the coast. Some of the finest limestone used in buildings around the world comes from Indiana. Even today, clams, corals, and various other shelled animals called mollusks continue to make shells in amazing variety. These shells continue to contribute to the calcium carbonate deposits in the ocean.

A large calcite crystal. Calcite and aragonite are actually two forms of the same chemical substance, calcium carbonate.

The calcium carbonate called chalk also comes from living things but not from the shells of mollusks. Instead, it is made up of tiny skeletons of ancient microscopic single-celled animals called foraminiferans. The texture of chalk is much finer than that of limestone because the skeletons forming chalk were much smaller than mollusk shells.

Limestone's Minerals

Calcium carbonate in the Earth's crust also exists as two different minerals, calcite and aragonite. These two minerals are metamorphic, meaning that they were changed, or metamorphosed, from the original sediments by heat and pressure. Calcite and aragonite have the same chemical composition but have different structures and densities.

The different crystal forms of calcium carbonate occur as a result of different conditions when they formed in the Earth's

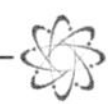

crust. Calcite was formed by higher temperatures but lower pressure, and aragonite formed with lower temperatures but higher pressure. Which one exists at any particular time or place depends on the pressure exerted on it from the rocks above and on the temperature of the rock layer. There is more calcite than aragonite because aragonite dissolves more easily than calcite.

Limestone isn't always pure calcium carbonate. In fact, some limestone that formed in water had some calcium replaced by the element magnesium (Mg, element #12). This change produces the mineral called dolomite—$CaMg(CO_3)_2$.

Limestone that has been compressed under great pressure and heated under the highest temperatures in the Earth's crust changes into the much harder—and more expensive—stone called marble. This very hard, beautiful stone is no longer sedimentary limestone. Instead, marble has become a coarser-grained crystalline rock that can be polished.

Although marble is primarily calcium carbonate, it often has impurities that give it wonderful streaks of color, such as green, blue, and red. Marble is rarely used as building stone. Instead, it is usually cut into slender sheets and fastened to another, cheaper material to make a lovely surface, or facade, for a building or fireplace.

Versatile Lime

When limestone is heated to almost 900°C (1,650°F), it changes character. Carbon dioxide, CO_2, is driven off by the heat and a white, chalky solid called lime, or quicklime, is left behind. This solid is calcium oxide, CaO.

$$2CaCO_3 + \text{heat} \rightarrow 2CaO + 2CO_2$$

The process of making lime—called calcination—is one of the main sources of carbon dioxide for industrial uses, and the lime itself has many uses.

A marble quarry in Georgia shows how marble and other types of limestone have formed deep layers in the Earth's crust. The rock can be cut into square pieces for use as building stone.

Lime can burn the skin, so it must be handled carefully. In centuries past, bodies buried without coffins sometimes had lime thrown on top of them to speed decomposition.

When CaO is mixed with water, the mixture gives off heat as the lime and the water combine in a process called slaking. The result is calcium hydroxide, $Ca(OH)_2$. Calcium hydroxide, which also often is called lime, is used as mortar. However, water vapor in air slowly changes the calcium hydroxide mortar back into calcium carbonate, which can crumble away. That's why the mortar holding brick or stone together often needs to be repaired. In the past, calcium hydroxide was used to make cucumbers crisp when they were turned into pickles. Now this process is not regarded as safe, but the chemical prevented other compounds from softening the cucumber.

Cement

Calcium oxide is mixed with clay and baked in a continuously rotating furnace to form a gray powder called cement. Mixed with water, cement can be used as a mortar to hold layers of bricks or stones together. It can also be shaped into many objects, such as inexpensive statues. Both of these processes must be carried out quickly because cement hardens rapidly.

A special kind of cement called portland cement is made by mixing calcium oxide with silicon and aluminum minerals. Named after the town of Portland in England, this cement hardens under water. Most cement used in the United States is portland cement.

Cement is usually mixed with water and gravel to make concrete. Perhaps you have seen a truck carrying concrete going down the road with its huge tumbler continually rotating. The tumbler is mixing the concrete and also keeping it from hardening until the material can be poured to make a sidewalk or a building foundation.

Gypsum

Like calcium carbonate, calcium sulfate ($CaSO_4$), called gypsum, formed over millions of years into thick layers, which can be mined. Gypsum is the white chalky material doctors used for centuries to form casts for broken arms and legs. As early as a thousand years ago, books recorded that gypsum was used to hold broken bones in place while they healed.

Because gypsum contains considerable water, it is sometimes called hydrous calcium sulfate and is written $CaSO_4 \cdot H_2O$. Natural calcium sulfate without any water is called anhydrite. Anhydrite is mainly used as a drying agent because it absorbs water easily and quickly.

Plaster of paris is powdered calcium sulfate that has been heated to about 190°C (375°F). This heat causes much of the water in calcium sulfate to evaporate. The resulting product is

Gypsum has long been used for plastering walls. Today, it comes as drywall, or plasterboard, which is simply fastened into place.

sold as a fine white powder. When water is added again, the plaster of paris can be molded and dried. Although a different substance is used today, plaster of paris was used for a long time to make molds of people's teeth for fitting braces or for making false teeth. Perhaps you've used plaster of paris to make casts of animal prints.

If calcium sulfate is heated to a higher temperature until all the water in it is driven off, it makes anhydrous (meaning "waterless") gypsum. For centuries, this powder has been mixed with sand and water to make plaster for use in coating the walls of buildings.

Until recently, plaster walls were the walls of choice in building homes and offices. Today, gypsum plaster is still used, but instead of coming as a powder, it comes as big sheets of already-made wall consisting of a layer of gypsum between thick sheets of paper. All a carpenter has to do it fasten the sheets in place. These sheets are called plasterboard or wallboard. They are also called drywall because they are used without mixing plaster powder and water. A special waterproof coating is put on drywall that is to be used in bathrooms, where the wall might get wet.

Spectacular Limestone

When the mineral calcite is transparent, it is called Iceland spar because it frequently has been found in the volcanic rock that makes up the island of Iceland. When the calcite is white, especially when it forms around springs, it is called travertine.

The beautiful travertine formed by Mammoth Hot Springs at Yellowstone National Park is calcite that was deposited by water.

Mammoth Hot Springs at Yellowstone National Park in Wyoming is one of the most spectacular displays of travertine. The hot water of the spring accumulates a great deal of calcium carbonate, $CaCO_3$, as it moves through underground rocks. Then, when the spring hits the air, the water cools so that it can no longer hold as much dissolved $CaCO_3$. The limestone precipitates (drops out) around the mouth of the spring, forming spectacular beds of white travertine rock.

Similar to travertine but much more crumbly is tufa. This is a porous limestone that also precipitates out of water. Mono Lake in California has a large number of spiky tufa rocks along its shores. In Turkey, tufa formed a system of tunnel-filled rocks where people have hidden over the years.

The movement of calcium compounds in water through rock can also produce some spectacular results in underground caves that have formed within limestone deposits. Rain can carry

Tufa rocks make fantasy shapes reflected in the moonlit water of Mono Lake in California.

carbon dioxide from the air down into the rock. The CO_2 and water dissolve some of the calcium carbonate in the limestone, forming calcium bicarbonate, $Ca(HCO_3)_2$. The water carrying the calcium bicarbonate gradually seeps through the rock until it forms drops of water on the roof of the cave.

Striking the air in the cave, the calcium bicarbonate changes back into calcium carbonate, or calcite. The drop of water

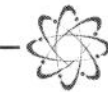

containing the calcite evaporates, and an infinitesimal bit of solid mineral is left behind. Later—perhaps much later—it is joined by another drip, and again the water evaporates, leaving more mineral behind.

Over many thousands of years, those tiny specks add up to hanging icicle-like chunks of calcite, called stalactites. Sometimes the drops fall to the floor of the cave before the water evaporates. When that happens, similar deposits called stalagmites build upward. In this way, over long eons of time, spectacular stalactites and stalagmites were formed at Mammoth Cave in Kentucky, Carlsbad Caverns in New Mexico, and in many other places around the world.

Stalactites and stalagmites in Carlsbad Caverns in New Mexico

MORE MILK, PLEASE

Many of the liquids we drink and the foods we eat contain calcium. This mineral is necessary for our bodies to grow and function properly. It is needed at all stages of our lives.

About 98 percent of the calcium in our bodies is in our bones. Only 1 percent is in our teeth. The remaining 1 percent is in our blood, muscles, and other soft tissues. But that last 1 percent controls what happens to the other 99 percent.

A developing baby needs calcium for its bones to form properly. A growing child also needs a regular supply of calcium for its bones to become long and sturdy and for strong teeth to form. An adult's body uses calcium to replenish the bones and teeth and for all the body's cells to function correctly. Finally, even a

Threads of fibrin trap red blood cells at the site of a wound. Calcium in the blood helps make blood clot.

person in old age requires plenty of calcium in the diet to keep bones from becoming brittle and easily broken.

Calcium in the Blood

The 1 percent of calcium that is not in the teeth or bones controls the movement of calcium ions into and out of those hard tissues. Calcium ions play an important role in many of the body's functions. They must be in the blood at all times.

Calcium ions in the blood are called into action when the body is cut. Calcium acts as a catalyst, which is a substance that must be present for a process to take place but is not actually part of the overall chemical reactions of the process. When the body is cut, proteins in the plasma (liquid part) of the blood change to a chemical called thrombin. Then the thrombin changes another chemical, fibrinogen, to fibrin. The threads of fibrin form a trap in which blood cells accumulate and plug the wound. Calcium helps the clotting happen. If there is not enough calcium in the blood, the thrombin needed for this complicated process to happen does not form, and bleeding continues.

Calcium ions also control the contraction and relaxation of muscle cells in the body. Calcium ions are stored in the body's muscle cells until the nervous system signals for calcium ions to be released. Once released into a muscle, these ions serve as a signal to the muscle fibrils (threads) to contract.

The level of calcium ions in the blood can affect the heart. People with high blood pressure often have to control the calcium ions in their blood. Increasing calcium and potassium in the diet can lower blood pressure, especially in combination with lowering fat intake. However, calcium can also narrow the blood vessels around the heart, making it more difficult for blood to flow through the blood vessels. This increases blood pressure. Special new medicines called calcium channel blockers may be used to keep blood vessels from narrowing.

The level of calcium ions in the blood can also be too low for the heart muscle to contract properly. If that happens, the body signals the parathyroid glands (located in the neck) to release some of their hormone, called parathormone. A hormone is a chemical "messenger" that regulates the actions of cells within the body. Parathormone travels through the body signaling special cells in the bone to release calcium ions that will repair the situation. Under more normal conditions, parathormone controls the absorption of calcium from food by the large intestine and from urine by the kidneys.

Ions in the Body

Calcium also controls the flow of other ions into and out of the cells through the cell walls. Other minerals needed by our bodies are potassium, magnesium, sodium, phosphorus (P, element #15), iron (Fe, #26), and sulfur (S, #16).

Sodium, potassium, magnesium, and iron atoms form cations, or positive ions. They can gain electrons. The negative ions, or anions, are all molecules that can lose electrons. They are made up of two or more elements.

All these ions, called electrolytes, control the movement of fluid into and out of cells. But electrolytes can disappear easily when a person loses too much liquid from the body. A person with a severe case of diarrhea or an athlete who exercises a lot may need to replenish the supplies of electrolytes in the body. An athlete can usually get enough electrolytes from sports drinks, but a sick person may have to go to the hospital and be given intravenous fluids containing electrolytes in order to replace what has been lost.

The elements that form electrolytes work together and thus must be absorbed together. It's not safe to build up one without balancing them all. For instance, calcium makes the heart muscle contract, moving the blood from one chamber to another or

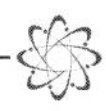

from the heart into the circulatory system. Magnesium makes the muscle relax after the contraction. And both sodium and potassium help to generate the electrical impulse that makes the contraction happen. All four of those minerals are equally important to the body.

Dynamic Bones

The bulk of the calcium in the body—98 percent—is in the bones. We think of bones much like the steel framework of a building—once formed, it doesn't change. And yet, our bones are dynamic. The materials that make them up are continually changing. In fact, bones are more like banks for calcium with the element being stored and taken out again as the body needs it.

The calcium in our bones occurs as apatite, any of several minerals containing calcium phosphate, $Ca_3(PO_4)_2$. The name apatite, which means "to deceive," is appropriate because several different minerals containing calcium phosphate are difficult to distinguish from each other. The most common calcium phosphate in bone is hydroxyapatite, $Ca_{10}(PO_4)_6(OH)_2$.

Apatite in the body doesn't just sit there unchanged. Instead, a continuous exchange of calcium ions (Ca^{2+}) and phosphate ions (PO_4^{3-}) from apatite goes on with ions from the blood.

Bone tissue is made up of two parts. First there is a web-like organic framework consisting of a protein called collagen. Filling in the holes in the web is the solid, fairly heavy inorganic calcium phosphate mineral. Without enough of this mineral, bones become soft and bendable. The second part of bone is calcite—crystalline calcium carbonate—which makes up the hard outer surface. Calcite in bone can replace itself if the bone is broken.

Bone is covered by a membrane called the periosteum. The inside of this membrane is where the special cells called osteoblasts form bone. (*Osteo* means "bone.") Osteoblasts are also found in a membrane in the hollow interior of the bone.

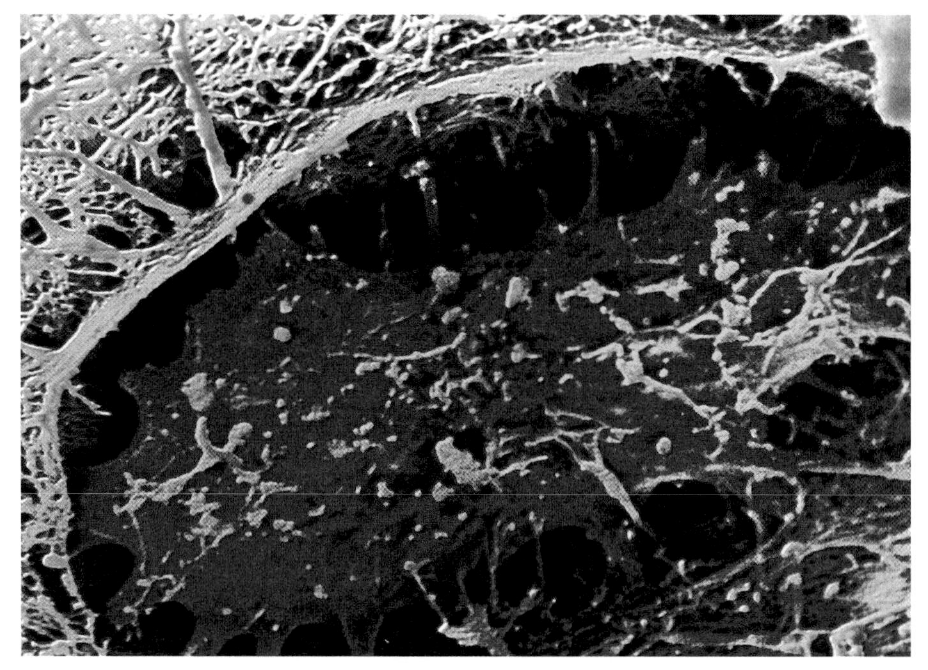

An osteoblast, colored by computer. This cell produces the collagen fibers (green) in which calcite crystals form.

Moving Calcium In and Out

When the body signals that calcium ions are required elsewhere in the body, special bone cells called osteoclasts remove calcium phosphate from nearby bone. Enzymes within the osteoclasts actually strip the calcium phosphate minerals from bone, forming tiny pits around the bone cell. The osteoclasts then secrete the ions into the bloodstream. This process is called resorption.

Once the osteoclasts have made tiny pits in the bone, another type of cell, the osteoblast, fills the pits up again. Up until about age twenty-five or thirty, a person's osteoblasts work

harder than the osteoclasts, and bone mass increases. Then, for a decade or two, the activity of the two types of cells is about equal. However, after that, the osteoclasts take over, and we begin to lose bone density.

Two hormones work together to keep the body's calcium in balance—parathormone, the hormone secreted by the parathyroid gland, and calcitonin, a hormone made by the thyroid gland. Parathormone increases the calcium in the bone tissue. Calcitonin slows the movement of calcium from the bones into the bloodstream. But this task becomes more and more difficult when people don't get enough calcium. Bones then become less dense. This is a condition called osteoporosis, which means "porous bones."

"Porous Bones"

Most vitamin and mineral deficiencies can be detected through blood testing. However, the lack of calcium in the diet does not show up in a blood test because the lack of calcium is in the bones, not in the blood. The deficiency in the bones may not show up for years—when it's too late to do anything about it.

Osteoporosis is a bone disease that usually occurs in adults, especially women. A person with osteoporosis is more likely to break bones than other people. In fact, bones in people with osteoporosis can break with just the simplest pressure. For example, a rib may break when a person with osteoporosis leans down over a chair arm to pick something up.

A broken rib doesn't sound too serious, but a broken hip can be very serious. A hip bone that has already lost a large part of its mineral content may shatter instead of just breaking in one spot. Such a break may take a long time to heal, if it ever does. Many elderly people never truly recover from broken hips. Almost 50 percent of the elderly people who break their hips end up living in nursing homes.

Everyone loses bone mass after about age 30, but women lose more than men, up to perhaps 40 percent. Taking supplemental calcium in large amounts won't return lost density, but it will stop further loss. The best treatment is one that starts early—in youth. Instead of giving up milk to drink lots of coffee, tea, and soda, young people should keep on drinking milk.

Exercise is needed at all stages of life to keep bones in good shape. Although it seems as if exercise would primarily benefit the muscles, it also helps keep bones healthy. Bones attached to unused muscles tend to waste away, or atrophy. Weight-bearing exercises are important for maintaining bone density.

Many women take prescription medicines that replace estrogen, the main female hormone. Estrogen has the added benefit of preventing bone density loss. Also, there are now prescription drugs that slow the activity of cells that contribute to bone loss.

In the Mouth

The hard part of a tooth—and the hardest substance in the body—is the enamel, or outer surface of the tooth. It is made of apatite, or calcium phosphate mineral. Tooth apatite continually gives up calcium ions (Ca^{2+}), phosphate ions (PO_4^{3-}), and hydroxide ions (OH^-). All three kinds of ions are replaced by ions in the food we eat. The continuous movement in and out of these ions makes teeth, like bone, living tissue.

Unfortunately, apatite is subject to being attacked by a film, called plaque, that is built up in our mouths by bacteria. Several kinds of bacteria accumulate in our mouths. Some of these bacteria have the ability to change plaque into an acid. That acid attacks both calcium ions and phosphate ions in the apatite. The result is weak spots that let other bacteria into the interior of the tooth, which will gradually be destroyed if left untreated.

If plaque is kept off the teeth, the enamel is continually reconstructed by the minerals in saliva. This fluid is produced by special glands in the mouth. It contains both calcium and phosphate ions, which keep the teeth from dissolving. But if too much sugar is in the diet or if the plaque is not regularly and adequately removed, the minerals in saliva cannot keep pace with the demineralization of the teeth, and decay results.

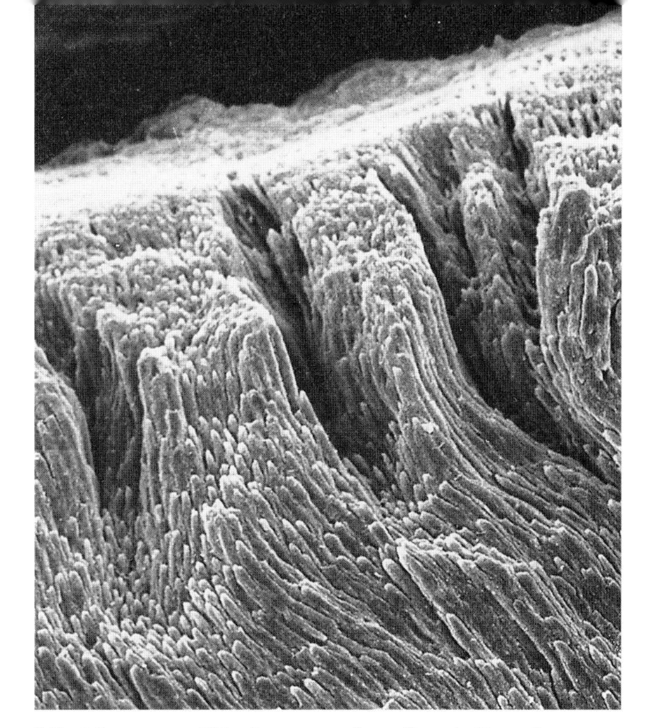

Highly magnified crystals of calcite that make up the enamel of teeth.

Caring for the Teeth

So what can you do about plaque? Exactly what you've always been told to do—brush regularly with something that will remove the plaque. A nice gentle toothpaste that leaves your breath feeling fresh is probably not enough. You need something that has an abrasive, or grinding, quality to it to break up the plaque without harming the enamel. One of the oldest abrasives for cleaning teeth is still one of the most useful—calcium carbonate.

Several calcium compounds are used as abrasives in various dental products. These include calcium carbonate, $CaCO_3$, calcium pyrophosphate, $Ca_2P_2O_7$, and tricalcium phosphate, $Ca_3(PO_4)_2$.

The chemicals generally called fluorides (which produce fluoride ions, F^-) change the chemistry of teeth. They slow the action of bacteria in making acids, and they also strengthen tooth enamel. Brushing with fluoridated toothpaste or drinking water containing fluorides helps protect teeth because—as noted above—the enamel is living tissue, with compounds moving in and out.

How Much Do You Need?

Clearly, we have to make sure that our bodies get all the calcium they need. But how much is that?

In 1997, the United States government changed its recommendations on the amount of calcium people should have in their diet. Previously, the recommended daily allowances (RDAs) were intended to prevent deficiency diseases. But, at the recommendation of the National Science Foundation, the government now suggests amounts of calcium in the diet that are intended to protect against diseases of all kinds.

Osteoporosis in old people, especially women, is most easily prevented from starting when people are teenagers. The RDA for both boys and girls from ages 9 to 18 is at least 1,300 milligrams of calcium every day. If this recommendation is followed, then, after age 50, when osteoporosis is likely to start, any deterioration of bone mass begins from a position of strength instead of weakness. The RDA for people over 50 is now 1,200 milligrams a day, so that as bone loses calcium it is replaced immediately.

Where Does It Come From?

An 8-ounce (0.24-l) glass of milk contains more than 200 milligrams of calcium. People who are concerned about the amount of fat in milk should know that there is just as much calcium in skim milk as there is in whole milk. So there's no reason to drop milk from your daily drinks.

Unfortunately, a rumor has been spread (perhaps by young people who didn't care for milk) that it is not good to drink milk before an athletic event. A major study found that rumor to be false and discovered that milk actually helps athletes perform.

Two slices of American cheese work even better than a glass of milk in providing calcium. And Swiss cheese contains more

Cheese (seen here being made) is a good source of calcium because it concentrates the mineral found in milk.

calcium than any other common cheese, despite the holes. Cheese is a concentrated form of calcium because the calcium ions (Ca^{2+}) in milk are mostly bound to the milk protein called casein. Casein clumps together to form a curd from which cheese is made.

Unfortunately, many people who want to lose weight tend to cut cheese out of their diet. Cheese is too valuable as a source of calcium to stop eating it completely. Cheese pizza is always a good way of getting calcium.

One serving of a food that is particularly high in calcium is a 85-gram (3-ounce) serving of sardines, which has 375 mg of calcium. Dark-green vegetables, such as broccoli, asparagus, beet and turnip greens, parsley, and okra, are also excellent sources of calcium. The outer dark-colored leaves of cabbage contain more calcium than the lighter-colored inner leaves.

Are You Getting It All?

Young children absorb up to 70 percent of the calcium they take in. However, an adult may absorb only about 30 percent. If

an adult is cutting down on calcium intake, he or she could be absorbing as little as 10 percent of the calcium required for a healthy body.

The body does not absorb all the calcium in milk either, but it absorbs more calcium from milk than from any other calcium-rich food, except broccoli and mustard greens. However, you'd have to eat five servings of broccoli to get as much calcium as you'd get from an 8-ounce (0.24-l) glass of milk.

The most important factor in the absorption of calcium is a chemical called vitamin D. Most vitamins have to be obtained from food, but vitamin D is also manufactured in our bodies when our skin is exposed to sunlight. Sunlight is so effective that any calcium you take within an hour or two of playing in the sun is going to be absorbed better than calcium taken at any other time, but exposure to the sun can be harmful.

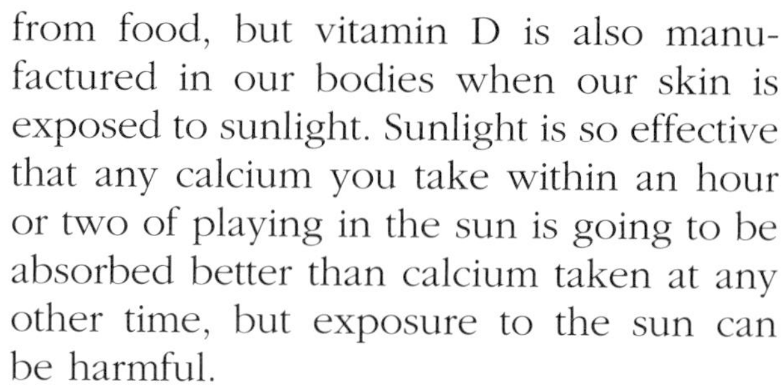

In cold places, people spend much of the year wearing clothing that covers them fully. A growing child in a cold climate can develop a disease called rickets from lack of calcium absorption. A child with rickets

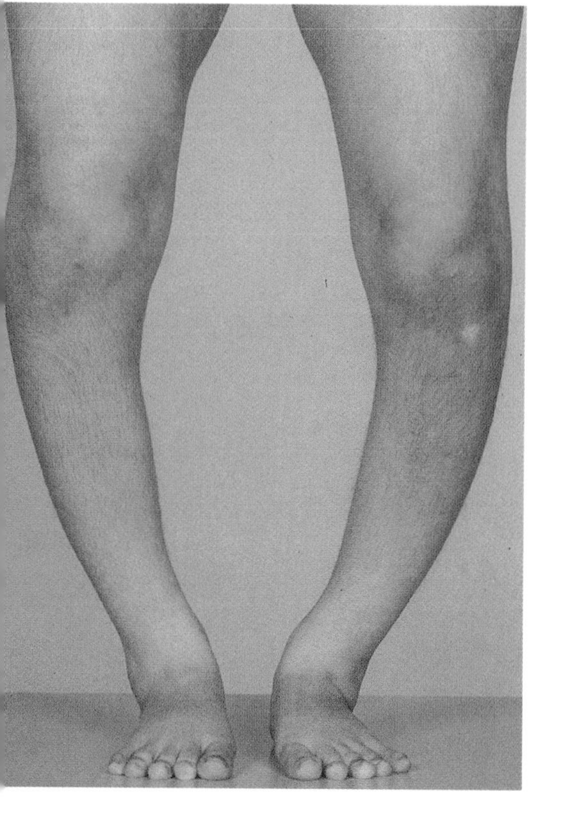

This photo shows a young person whose legs bow out from the condition called rickets. This disease is caused by lack of calcium in the diet during childhood. Rickets is not as common as it once was. Drinking milk or taking calcium supplements can help a person avoid calcium deficiencies.

has legs that bow outward severely. Infants who don't get enough milk in the early months of life may also develop rickets.

Today, vitamin D is added regularly to milk. Before this practice was started, children were often given doses of cod liver oil to get their supply of vitamin D. Cod liver oil is literally an oil made from the liver of the codfish. It was a source of vitamins before they were made in laboratories.

Too much vitamin D can cause the body to absorb too much calcium, allowing calcium deposits to form in parts of the body where they don't belong, such as in the body's soft tissues. If, for example, calcium accumulates in the heart or kidneys, it can cause death by blocking the flow of fluids. If it collects in joints, it can cause painful arthritis. Less seriously, it can cause hard lumps to form in the skin.

The calcium in some foods is prevented from helping the body because the foods contain a chemical called oxalic acid. Oxalic acid binds with calcium to form calcium oxalate, which is useless to the body. Spinach is one of the foods that contains oxalic acid, but it's so rich in iron and various vitamins that it shouldn't be left out of the diet.

Also, people who get too much protein or too much salt in their diet tend to lose a lot of calcium in their urine. Calcium absorption also can be slowed by stress, smoking, and the caffeine in coffee, tea, cola drinks, or chocolate.

Switching from milk to soda pop can double the problem of not having enough calcium in your body. Your body must have magnesium to absorb calcium. Soft drinks tend to decrease the amount of magnesium in the body. Even if you're also getting lots of calcium, it might not be absorbed properly if you drink a lot of soda pop. Also, some soft drinks are made with phosphates. One serving of soft drink may provide half a day's requirement of phosphorus, but the soft drink contains no calcium, and phosphorus without calcium is not good for the body.

Some people take calcium supplements made from oyster shells, thinking that such a natural source is best. But this form of calcium is usually not absorbed by the body. The most absorbable forms of calcium supplements are calcium citrate and calcium aspartate.

Limestone in the Stomach

The main ingredient in many antacids—chemicals that neutralize excess acid in the stomach—is calcium carbonate, which, of course, is limestone. The calcium carbonate in antacids breaks down in the stomach into calcium ions, Ca^{2+}, and carbonate ions, CO_3^{2-}. The acid that is in the stomach gives up negative hydrogen ions (H–). Relief occurs when the two kinds of ions react, making two harmless products, water and carbon dioxide:

$$CO_3^{2-} + 2H^- \rightarrow H_2O + CO_2$$

The leftover calcium ions enter the bloodstream and make their way into bone or other cells. That is why the makers of calcium carbonate antacids can advertise that their products are good sources of calcium.

A person who suffers from frequent heartburn and takes calcium-containing antacids in addition to drinking lots of milk may be prone to a condition called milk-alkali syndrome. One of the unfortunate effects of this condition is that it causes the stomach to produce even more acid than it would otherwise.

Other major kinds of antacids contain magnesium or aluminum. Those elements actually block the absorption of calcium. Therefore, calcium-containing antacids are better for people who might be prone to bone problems.

An oversupply of calcium can also come from an overactive parathyroid gland, as well as various tumors. This condition, called hypercalcemia, can cause nausea and vomiting, high blood pressure, and kidney stones.

OF WATER AND EGGS

Calcium is a natural part of our environment. All living things need it—but not necessarily from the places where nature puts it. Science and technology have played a role in changing what happens to the calcium in the environment. Sometimes that role has had a good effect, but sometimes it has been very harmful.

Calcium in Our Water

As you saw earlier, calcium ions, Ca^{2+}, are picked up from the calcium carbonate in limestone and dolomite as water flows across those rocks. Water containing calcium and magnesium ions is considered "hard."

Hard water causes soap to form a scum on clothing instead of rinsing away. In a bathtub, hard water can make soap form a gray ring around the tub that is difficult to rinse away.

The water that has collected within this limestone deposit is high in calcium.

An electronic system is now being used to clean the calcium carbonate deposits, or scale (see top pipe) out of municipal water pipes, saving considerable money and labor.

In the past, sodium carbonate, Na_2CO_3, also called washing soda, was added to wash water to keep scum from forming on clothing. The sodium carbonate reacted with calcium ions to form calcium carbonate:

$$Ca^{2+} + Na_2CO_3 \rightarrow 2Na^{+} + CaCO_3$$

Neither the new calcium compound nor the new sodium ions formed react with soap. They both rinse away. Today, detergents contain chemicals that make water rinse away cleanly.

When hard water is heated, some of the water evaporates. The smaller amount of water left can't hold as much calcium carbonate as the larger amount of water did, so some of the calcium carbonate precipitates out. That precipitated calcium carbonate can build up inside water pipes, steam irons, and teakettles, where it eventually plugs up the works.

To prevent this buildup, home water is often "softened" (the dissolved mineral is removed) in devices that exchange less

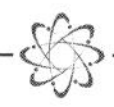

troublesome sodium ions for the problem-causing calcium ions. Calcium and magnesium ions that make water "hard" are captured by special chemicals and exchanged for sodium. The sodium ions come from dissolved salt, or sodium chloride (NaCl). Chlorine is C, element #17.

How Birds' Eggs Form

Natural calcium-containing water is the source of calcium for many living things. For birds, the calcium in the water they drink contributes to making the eggs they lay.

A bird's egg forms inside its body. Even before the shell forms, the egg has a rounded shape held in by a fairly tough membrane. The egg makes its way into the bird's uterus, where little granules of calcium carbonate are deposited onto the egg membrane. Several layers of shell are formed over a period of many hours, with different structures and molecules in each layer, though all the layers contain calcium. The final, outer layer of the eggshell is made of crystals of calcite.

Because of their commercial importance, hen's eggs are studied most. It is known that the calcium content of the blood of a hen that's laying eggs is two to three times higher than that of a hen that isn't laying eggs. The calcium makes its way through the uterus, onto the egg. If a bird lays eggs that have a color (like the blue of a robin's egg), that color comes from a chemical in the bird's red blood corpuscles.

In addition to getting calcium from the water they drink, hens in a poultry yard are given a pebble-like gravel containing lots of calcium with their food. Traditionally, the gravel contains chunks of oyster shell or limestone. These substances both readily dissolve in the hen's body, and the calcium compounds they contain are used by the hen's body to make eggs. If the calcium source is taken away, a hen's body gradually uses up its stored calcium and the hen stops laying eggs.

A variety of things in the environment can interfere with a bird's use of calcium in making eggs. One chemical had unforeseen and devastating effects on birds in the wild.

The Environmental Story

An insecticide called DDT was found in the 1940s to be amazingly effective against disease-carrying pests. DDT—chemically dichlorodiphenyltrichloroethane ($C_{14}H_9C_{15}$)—was soon used around the world, especially in places where mosquitoes spread the killer diseases of malaria and typhus. The use of the insecticide was soon taken for granted because it was so

The sight of a bald eagle in her nest almost became a thing of the past because DDT in the food chain made the eggshells of most birds of prey too weak to survive the incubation period.

effective. In the United States, DDT was regularly sprayed in virtually every town to eliminate a disease that was killing elm trees by the millions.

It wasn't until 1962 that naturalist Rachel Carson brought to the attention of the public the idea that DDT might be doing more harm than good. She said it was killing birds and other wildlife. The title of her book, *Silent Spring,* told people that our birds were disappearing. Many birds, especially birds of prey such as eagles and peregrine falcons, were not breeding fully anymore. They didn't lay as many eggs as they previously had, and the eggs they did lay had thin shells. The embryos inside the shells rarely lived long enough to develop and hatch. The number of birds in each generation was shrinking.

Many scientists who had supported the spraying program said that Carson, who was "an oceanographer without a Ph.D.," didn't know what she was talking about. It was another ten years before the Environmental Protection Agency banned the use of DDT in the United States.

DDT and other pesticides like it contain chlorine and remain in the environment for a long time. They don't break down into harmless compounds as some pesticides do. Small animals eat the plants with DDT on them. Then other animals eat those smaller animals. Finally, the raptors, or birds of prey, eat those animals. At each stage, the DDT accumulates in the animals, until finally, the predatory raptors get a full magnified load of the chemical.

The reaction of the eggshell-making mechanisms to the presence of DDT is a complicated one. But apparently the chemical prevents the action of an enzyme that plays a role in making the bicarbonate ion (HCO_3^-), which is part of the process of making calcium carbonate. The birds at the top of their food chain were laying eggs with shells that were dangerously thin.

The ban on DDT use and work by naturalists over thirty years have rebuilt the populations of eagles and hawks. In 1994,

The tiny animals called polyps (above) make shells that fuse to the shells of other polyps, gradually forming coral reefs. A coral reef (shown at left in the bottom half of the picture) serves as the foundation for many different kinds of living things in warm seas.

the bald eagle, America's symbol, was taken off the endangered species list. But it should never have been endangered in the first place. Rachel Carson's *Silent Spring* made people realize that the things humans do can harm the whole of Earth's natural world.

Calcium Reefs

Some of the most spectacular structures of calcium material are the coral reefs that make wonderful underwater settings for many plants and animals, usually in warm, shallow seas. A coral is actually a tiny animal called a polyp that makes a hard outer skeleton of calcium carbonate. Many corals live together, their skeletons fusing to each other. Generation after generation of the

animals build new layers of coral, gradually forming the great stone-like structures called reefs.

Contributing to the construction of reefs are special types of primitive plants called algae. Some members of the algae family take in calcium carbonate from the ocean and build up a crystalline structure similar to the coral's. As these calcareous algae die, they leave behind their calcium carbonate, which contributes to the structure of the reefs.

Reefs serve as home—or at least as a safety zone—for many animals. Some of the most beautiful tropical fish are found among coral reefs, and that is one of the reasons that the great reefs of our planet are in danger. Divers, eager to see the beautiful fish, can easily smash coral beds, destroying them.

Not all reefs have the same shape. An atoll is a circular reef that forms around and encloses a body of water called a lagoon. Other reefs called barrier reefs lie parallel to the shore. The greatest barrier reef known—indeed, probably the largest structure made by living creatures—is the Great Barrier Reef. It lies along the eastern coast of Australia and is more than 2,010 kilometers (1,250 mi) long. That calcium structure, built up over many thousands of years, is so large that it can be seen by an astronaut in outer space.

Calcium in Soil

Because there is so much limestone in our world, most soil usually does not need added calcium to make it good for growing plants. Some plants grow best in soil that is alkaline while others prefer an acid soil. Calcium is sometimes added to soil to change a soil's acidity.

Calcium hydroxide—the "slaked" lime made from calcium oxide—may be added to soil that is too acidic for certain plants. When the $Ca(OH)_2$ mixes with water, it breaks into hydroxide ions (OH^-) and calcium ions (Ca^{2+}). The hydroxide ions react with hydrogen ions in the acid content of the soil to form water. Soil particles have, by nature, a negative charge, so they attract the positive calcium ions. The positive calcium ions make the soil particles clump together so that water moves through it freely, rather than getting stopped by a mass of clay.

A plant needs calcium in order to build new cell walls. A soil with too much calcium may keep plants from growing properly and prevent other elements from being taken up.

Many trees, as well as some other plants such as orchids, do not grow properly unless they have a certain fungus growing within their roots. This relationship between a higher plant, such as a tree, and a primitive plant (a fungus) is called a mycorrhiza (*myco* means "fungus" and *rhiza* means "root"). The fungus takes in calcium and other minerals that the tree roots can't absorb directly. Then the tree can take the minerals from the fungus.

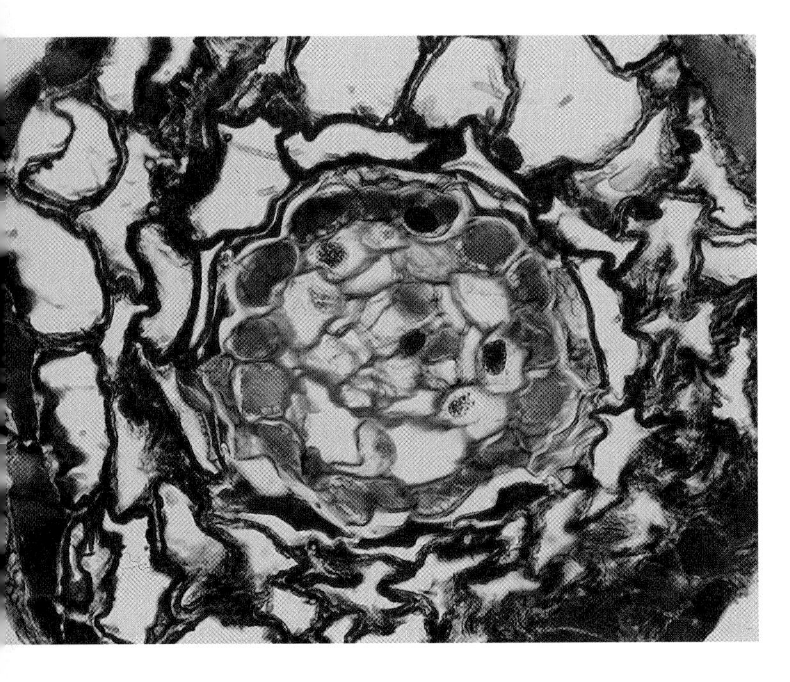

A mycorrhiza is shown in this cross section of a plant root. The red threads are the fungus, which can absorb calcium and transfer it into the root, which is the center circular part.

ART AND INDUSTRY

Because limestone and many products that come from it are so common in our world, they have been used for many different purposes. Fine-quality lime, for example, has been used for centuries by artists to make wall paintings called frescoes.

The word fresco means "fresh." A wall is freshly coated with a thin layer of plaster, made with lime and water. Then, while the coating is still wet, the artist paints a picture on it. As the lime dries and hardens, the picture solidifies, becoming part of the wall.

This kind of painting was first known in ancient Greece, but it reached its glory during the 1500s in Italy. Michelangelo's great paintings in the Sistine Chapel in Rome are frescoes still admired today.

A fresco painted on a ceiling of the U.S. Capitol in Washington, D.C.

Michelangelo also used calcium in the form of marble for sculpting statues. The most prized marble is Carrara marble, which comes from a mine in Italy. Michelangelo's sculptures of Carrara marble have remained beautiful for hundreds of years. The scarce white Carrara marble is still used today by sculptors for their finest statues.

There is one problem with marble statues and walls. It's a problem that was not realized until the twentieth century. Coal that contains a great deal of sulfur emits sulfur dioxide, SO_2, as it burns. Refining petroleum also yields sulfur dioxide. Sulfur dioxide reacts with oxygen in the air to form sulfur trioxide, SO_3. It, in turn, reacts with water vapor to form sulfuric acid.

$$2SO_2 + O_2 \rightarrow 2SO_3$$

$$SO_3 + H_2O \rightarrow H_2SO_4$$

Acid droplets in the air can fall with rain, striking marble facades or statues. The sulfuric acid reacts with calcium carbonate in the marble to form calcium bicarbonate, $Ca(HCO_3)_2$. This compound actually fizzes, or effervesces, like a soda. The escaping carbon dioxide carries away some of the material in the marble, gradually destroying ancient statues.

In order to limit such air-pollution damage, factories that burn high-sulfur coal usually have devices called scrubbers on their chimneys. Scrubbers remove sulfur from emissions. A scrubber often contains calcium oxide, which reacts with the sulfur dioxide to form calcium sulfate ($CaSO_4$), which is gypsum. This gypsum is recovered and used in making drywall. So, instead of destroying things, the sulfur is now put to good use.

The ancient Egyptians used wet gypsum for sculpting. When calcium sulfate is mixed with water, it gives off a great deal of heat, making the water evaporate quickly as the mixture dries. Despite the fact that the artist had to work quickly, the Greek

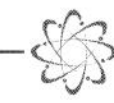

writer Theophrastus described gypsum as "superior to all other things, for making images."

A very fine-grained gypsum is called alabaster. It has a translucent gleam, rather like marble. Alabaster is easier to carve than marble, however, and so it has long been used to make ornamental objects such as vases or figurines. Unlike regular gypsum, which is very common, alabaster is found in only a few places around the world.

Calcium Chloride Against Cold Temperatures

Crystals of calcium chloride, $CaCl_2$, occur naturally in deposits called brines and are also produced as a by-product of several industrial processes. At least 40 percent of the calcium chloride manufactured is spread on icy roads in winter. It lowers the freezing point of water to –55°C (–67°F) so that already frozen ice will melt. However, there is growing concern about the environmental effects of spreading calcium chloride and sodium chloride on the nation's roads. The chlorine in both those compounds can

Large "scrubbers" mounted at the base of the chimney of a coal-burning power plant remove harmful sulfur from the smoke by converting it into calcium sulfate.

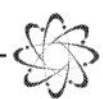

seep into our water supplies and make them too salty to drink. The calcium chloride crystals also have the ability to absorb moisture from the air. This makes the chemical useful as a drying agent, or desiccant. Small packets of calcium chloride are sometimes put into packages of merchandise to prevent moisture from harming the product before it is sold.

Calcium chloride is also used in making instant "hot packs" to provide heat for an athletic injury. Hot packs are plastic packets containing both calcium chloride crystals and an inner pouch of water. When the inner pouch is broken, the calcium chloride crystals dissolve in the water and give off enough heat to raise the temperature of the water. That heat is sufficient to ease the pain of a pulled muscle. A reaction that gives off heat is called an exothermic reaction.

Calcium chloride can also be used in batteries intended for use at cold temperatures. Usually the electrolyte (the ion-producing material) in a battery is zinc chloride. (Zinc is Zn, element #30.) But zinc ions stop moving below 18°C (0°F). Calcium ions keep electricity flowing down to a chilly –40°C (–40°F).

Using Those Valence Electrons

Because calcium has only two electrons in its outer shell, it readily reacts with other chemicals. Various industries take advantage of calcium's readiness to give up electrons by combining it with other elements to purify different materials.

As we have seen, some types of coal naturally have sulfur in them. Coal is used to heat iron in the manufacture of steel, but the sulfur content of the coal is not wanted in the steel. When calcium is added to the mix, it reacts with the sulfur to make calcium sulfide, which precipitates out of the mixture. In the same way, bismuth (Bi, element #83) can be removed from molten lead (Pb, #82) and copper (Cu, #29) when those metals are smelted from their ores.

The heat generated by burning acetylene gas is used to cut and weld steel.

Calcium Carbide

Calcium carbide (CaC_2) is a colorless, transparent solid that reacts with water to give off a gas called acetylene in the following process:

$$CaC_2 + 2H_2O \rightarrow Ca(OH)_2 + C_2H_2$$

This is not a quiet reaction. It is explosive, producing considerable acetylene gas and very bright light. The process was used in the 1800s to provide a light source in mines. The light was made by dripping water on small chunks of calcium carbide. The first headlights for automobiles were small acetylene lights, but these lights were not safe.

Today, acetylene is one of many products that come from the "cracking" or breaking apart of petroleum into its many parts, or fractions. Acetylene gas is captured and kept under pressure for use in high-temperature industrial welding. When combined with oxygen, the flame of an acetylene torch can reach a temperature of 3300°C (about 6000°F). This is one of the hottest flames that is readily available for industry, and it is most often used for cutting steel. Acetylene is also the basic ingredient in many complex polymers for plastics, vinyl chloride, and other organic compounds.

In the past, calcium carbide had another important use. It was used to take nitrogen from the air and incorporate the nitrogen into a compound that was used as a fertilizer. Such a process is called "fixing" because atmospheric nitrogen can't combine normally with other elements.

In another process, nitrogen is separated from the other gases in air by compressing the air into a liquid. When the liquid is heated, the different components evaporate at different temperatures. This means they can be separated out at different stages of the process. This process is called fractional distillation, because each gas is a different part, or fraction, of the liquid air.

The nitrogen produced by this process has been used to react with calcium carbide at a high temperature, making calcium cyanamide, CaNCN. This compound was then used directly as a fertilizer or made into ammonia, which is an even more efficient fertilizer. Today, ammonia is made from atmospheric nitrogen by directly combining nitrogen and hydrogen at very high temperatures under very high pressures.

Calcium in the Food Industry

Because of the demonstrated need for calcium in our diets, compounds containing calcium are being added to an increasing number of prepared foods. However, calcium compounds also serve other purposes. Several are put into commercially baked bread, primarily as preservatives. Calcium proprionate, for example, is added to bread to prevent mold from growing.

Calcium stearoyl lactylate is an additive used in the kind of whipped cream that comes in squirt cans. That calcium compound increases the volume of the fluid as it emerges from the can, making the whipped cream fluffy.

Animal bones are sometimes burned to create bone char, a granular combination of calcium and carbon. Bone char is used by the sugar-making industry to filter the unwanted color out of liquid sugar before it crystallizes.

A CALCIUM CATALOG

Pearly Crystals

Beautiful abalone shells are often used in jewelry for their pearly look. Abalone is a combination of calcite and aragonite, which are two forms of calcium carbonate that differ in the way their atoms are arranged. Scientists at the University of California at Santa Barbara have found that when an abalone is forming its shell, a layer of calcite is deposited first, and then a layer of aragonite. The calcite and aragonite form crystals in the pores of the proteins in the abalone. The proteins are directed by a genetic "switch" that controls the change from one kind of calcium carbonate to the other.

The shells of mollusks (shelled animals such as snails, oysters, and clams) are secreted by a body part called the mantle. The mantle of some mollusks

Part of a nautilus's elaborate shell has been cut away to show the pearly interior of calcite.

makes an inner layer in their shell with a pink iridescent coloring. Such a deposit is called mother-of-pearl, or nacre. It consists of alternating layers of calcium carbonate and a material called conchiolin.

The prize producers of nacre are certain oysters that react to an accidental grain of sand under their shell by surrounding the sand with layer after layer of nacre. The objects thus formed are called pearls. They are generally irregular in shape and are attached to the oyster's inner shell. But once in a great while the pearl separates from the shell and is perfectly spherical in shape. These pearls are highly prized.

People now create "cultured" pearls by maintaining huge beds, or farms, of oysters and inserting grains of sand in each mollusk's shell. The number of pearls that develop is far greater than when nature lets the grain in by accident, so cultured pearls are less valuable than natural pearls.

Look for the Holes

Shells lie on beaches all over the world. They are the remains of mollusks that have died. The soft mollusk bodies have decomposed, leaving their shells behind them. If you've ever investigated those shells, you

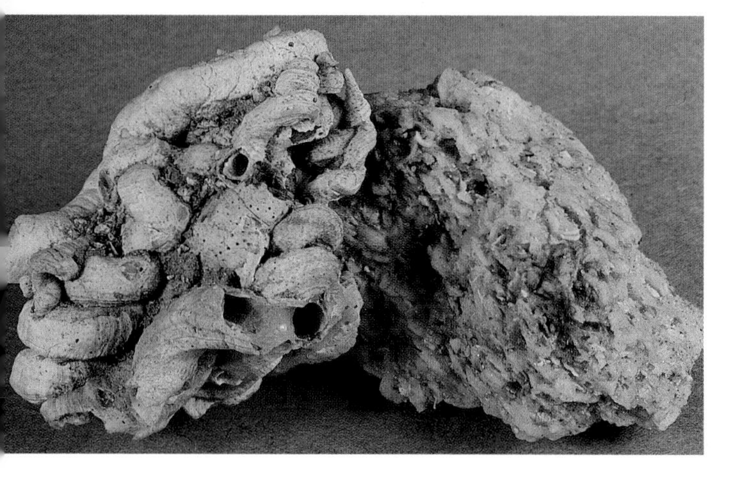

The shells of tiny clams called coquinas are visible in this limestone called coquina limestone. Coquinas are still found under the sand on beaches all over the world.

may have noticed that many have small holes in them, rarely more than 1.3 centimeters (0.5 in) in diameter. This indicates that the soft body of the creature in the shell probably served as food for another shelled animal—the conch. Conchs make the large coiled pink shells in which people think they can "hear the sea."

The holes in a conch's prey are formed when the conch attaches itself to a (usually) smaller shellfish. The conch then produces hydrochloric acid that dissolves the smaller shellfish's shell at the spot where the conch is clinging. The acid eats away a hole, and the conch can suck the prey out of its shell for a meal.

The Shell Eaters

Some other animals eat their own shells. Crabs, lobsters, and related animals grow by casting off their shells and growing new, bigger ones. After shedding its shell, the "naked" animal lies in hiding, helpless, until its new shell begins to harden. These invertebrate animals don't waste their old shells. They eat them, absorbing the calcium from them and using the calcium to grow new shells.

Calcium and Radioactivity

Elements in the same column of the Periodic Table can often substitute for each other in many chemical reactions. This can be a problem when a radioactive element is involved. A radioactive element emits particles from its nucleus, which can be dangerous to living creatures. Both radium (Ra, element #88) and some isotopes of strontium (Sr, #38) can replace calcium in the bones of people exposed to these radioactive elements.

Only after the deaths of some people who worked with radium was it realized that this element can be harmful. The radium gradually replaced some calcium in the bone, damaging the bone marrow. The marrow is where our infection-fighting white blood cells are manufactured. Radium in the bone marrow can cause leukemia, or cancer of the blood. In leukemia, the white-cell manufacturing process does not work properly.

Marie Curie, who discovered radium with her husband, died of leukemia. So did her daughter, Irène Joliot-Curie, who had also worked with the element.

Unfortunately, the radioactive isotope strontium-90 can also replace calcium in bones. When a nuclear reactor at Chernobyl in Ukraine had a major fire in 1986, a great deal of strontium-90 was released into the atmosphere. It fell onto the grasslands for many miles surrounding Chernobyl. Cattle ate the grass, and the radioactive isotope showed up in their milk, replacing calcium. Strontium-90 lasts up to 28 years in the environment, so large parts of Ukraine are now useless for agriculture.

In the Limelight

Before electricity was used to put spotlights on actors on a stage, calcium oxide, or lime, was used to make a bright light. The white lime powder was exposed to a very hot gas flame. The flame flared with an intense white light that could be beamed onto the stage. The white light that was produced this way was called Drummond's limelight. Would-be actors and entertainers still yearn to be "in the limelight."

Another Light

The mineral called fluorspar is calcium fluoride, CaF_2. If a light is shone on fluorspar varieties, the mineral re-emits light after the original light source is removed. This quality of fluorspar gave the name fluorescence to any process involving the emission of light from a source that does not itself make the light. The chemicals used in a fluorescent lamp absorb invisible ultraviolet light and re-emit it as visible light. However, fluorescent lamps have nothing to do with either calcium or fluorine.

Those Yellow Edges

You may have come across an old book with pages that have yellowish edges. The yellowish edges form because there is considerable acid in the paper, and that acid reacts with air, changing the color of the paper. The process can be stopped by dipping the book into a solution of calcium bicarbonate. This compound neutralizes the acidity of the paper, preventing further damage. Many books are now printed on acid-free paper that will last a lot longer than paper with acid.

Naming the Past

The chalk found in the Earth's crust, as in the white cliffs of Dover, is finely grained calcium carbonate. It was laid down when ancient seas were inhabited by billions of tiny shelled creatures called foraminifera. This period of time started about 140 million years ago and lasted about 75 million years. Geologists give the name cretaceous to this period. The word comes from *creta,* a Latin word meaning "chalk-like." The white cliffs of Dover were formed during the Cretaceous Period.

Calcium carbonate, or limestone, laid down in the Cretaceous Period. The photo shows the limestone both before (right) and after (left) it has been cleaned.

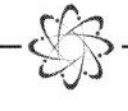

Calcium in Brief

Name: calcium, from the Latin word *calx,* meaning "lime"
Symbol: Ca
Isolated by: Humphry Davy in 1808
Atomic number: 20
Atomic weight: 40.08
Electrons in shells: 2, 8, 8, 2
Group: 2; other elements in Group 2 include beryllium, magnesium, strontium, barium, and radium
Usual characteristics: hard silvery metal
Density (mass per unit volume): 1.55 g/cm^3
Melting point (freezing point): 845°C (1,550°F)
Boiling point (liquefaction point): 1,420°C (2,590°F)
Abundance:
Universe: One of many elements totaling 1.1%
Earth: 7th most abundant (1.1%)
Earth's crust: 5th most abundant element at 3.4%
Earth's atmosphere: None
Human body: 5th most abundant element, making up 1.5% to 2% of the body by weight, virtually all of which is in the bones and teeth
Stable isotopes (calcium atoms with different numbers of neutrons in their nuclei): six stable isotopes occur in nature: Ca-40 (which makes up 96.97% of all calcium), 42 (0.65%), 43 (0.14%), 44 (the second most abundant, with 2.09%), 46 (0.004%), and 48 (0.19%)
Radioactive isotopes: Ca-36, 37, 38, 39, 41, 45, 47, 49, 50, and 51

Glossary

acid: definitions vary, but basically it is a corrosive substance that gives up a positive hydrogen ion (H+), equal to a proton when dissolved in water; indicates less than 7 on the pH scale because of its large number of hydrogen ions

alkali: a substance, such as an hydroxide or carbonate of an alkali metal, that when dissolved in water causes an increase in the hydroxide ion (OH-) concentration, forming a basic solution.

anion: an ion with a negative charge

atom: the smallest amount of an element that exhibits the properties of the element, consisting of protons, electrons, and (usually) neutrons

base: a substance that accepts a hydrogen ion (H+) when dissolved in water; indicates higher than 7 on the pH scale because of its small number of hydrogen ions

boiling point: the temperature at which a liquid at normal pressure evaporates into a gas, or a solid changes directly (sublimes) into a gas

bond: the attractive force linking atoms together in a molecule or crystal

catalyst: a substance that causes or speeds a chemical reaction without itself being consumed in the reaction

cation: an ion with a positive charge

chemical reaction: a transformation or change in a substance involving the electrons of the chemical elements making up the substance

compound: a substance formed by two or more chemical elements bound together by chemical means

covalent bond: a link between two atoms made by the atoms sharing electrons

crystal: a solid substance in which the atoms are arranged in three-dimensional patterns that create smooth outer surfaces, or faces

decompose: to break down a substance into its components

density: the amount of material in a given volume, or space; mass per unit volume; often stated as grams per cubic centimeter (g/cm^3)

dissolve: to spread evenly throughout the volume of another substance

distillation: the process in which a liquid is heated until it evaporates and the gas is collected and condensed back into a liquid in another container; often used to separate mixtures into their different components

electrode: a device such as a metal plate that conducts electrons into or out of a solution or battery

electrolysis: the decomposition of a substance by electricity

electrolyte: a substance that when dissolved in water or when liquefied conducts electricity

element: a substance that cannot be split chemically into simpler substances that maintain the same characteristics. Each of the 103 naturally occurring chemical elements is made up of atoms of the same kind.

evaporate: to change from a liquid to a gas

gas: a state of matter in which the atoms or molecules move freely, matching the shape and volume of the container holding it

group: a vertical column in the Periodic Table, with each element having similar physical and chemical characteristics; also called chemical family

half-life: the period of time required for half of a radioactive element to decay

hormone: any of various secretions of the endocrine glands that control different functions of the body, especially at the cellular level

ion: an atom or molecule that has acquired an electric charge by gaining or losing one or more electrons

ionic bond: a link between two atoms made by one atom taking one or more electrons from the other, giving the two atoms opposite electrical charges, which holds them together

isotope: an atom with a different number of neutrons in its nucleus from other atoms of the same element

mass number: the total of protons and neutrons in the nucleus of an atom

melting point: the temperature at which a solid becomes a liquid

metal: a chemical element that conducts electricity, usually shines, or reflects light, is dense, and can be shaped. About three-quarters of the naturally occurring elements are metals.

metalloid: a chemical element that has some characteristics of a metal and some of a nonmetal; includes some elements in Groups 13 through 17 in the Periodic Table

molecule: the smallest amount of a substance that has the characteristics of the substance and usually consists of two or more atoms

monomer: a molecule that can be linked to many other identical molecules to make a polymer

neutral: 1) having neither acidic nor basic properties; 2) having no electrical charge

neutron: a subatomic particle within the nucleus of all atoms except hydrogen; has no electric charge

nonmetal: a chemical element that does not conduct electricity, is not dense, and is too brittle to be worked. Nonmetals easily form ions, and they include some elements in Groups 14 through 17 and all of Group 18 in the Periodic Table.

nucleus: 1) the central part of an atom, which has a positive electrical charge from its one or more protons; the nuclei of all atoms except hydrogen also include electrically neutral neutrons; 2) the central portion of most living cells, which controls the activities of the cells and contains the genetic material

oxidation: the loss of electrons during a chemical reaction; need not necessarily involve the element oxygen

pH: a measure of the acidity of a substance, on a scale of 0 to 14, with 7 being neutral. pH stands for "potential of hydrogen."

pressure: the force exerted by an object divided by the area over which the force is exerted. The air at sea level exerts a pressure, called atmospheric pressure, of 14.7 pounds per square inch (1013 millibars).

protein: a complex biological chemical made by the linking of many amino acids

proton: a subatomic particle within the nucleus of all atoms; has a positive electric charge

radical: an atom or molecule that contains an unpaired electron

radioactive: of an atom, spontaneously emitting high-energy particles

reduction: the gain of electrons, which occurs in conjunction with oxidation

respiration: the process of taking in oxygen and giving off carbon dioxide

salt: any compound that, with water, results from the neutralization of an acid by a base. In common usage, sodium chloride (table salt)

shell: a region surrounding the nucleus of an atom in which one or more electrons can occur. The inner shell can hold a maximum of two electrons; others may hold eight or more. If an atom's outer, or valence, shell does not hold its maximum number of electrons, the atom is subject to chemical reactions.

solid: a state of matter in which the shape of the collection of atoms or molecules does not depend on the container

solution: a mixture in which one substance is evenly distributed throughout another

sublime: to change directly from a solid to a gas without becoming a liquid first

synthetic: created in a laboratory instead of occurring naturally

triple bond: the sharing of three pairs of electrons between two atoms in a molecule

ultraviolet: electromagnetic radiation which has a wavelength shorter than visible light

valence electron: an electron located in the outer shell of an atom, available to participate in chemical reactions

vitamin: any of several organic substances, usually obtainable from a balanced diet, that the human body needs for specific physiological processes to take place

For Further Information

BOOKS

Atkins, P. W. *The Periodic Kingdom: A Journey into the Land of the Chemical Elements.* NY: Basic Books, 1995

Heiserman, David L. *Exploring Chemical Elements and Their Compounds,* Blue Ridge Summit, PA: Tab Books, 1992

Hoffman, Roald, and Vivian Torrence. *Chemistry Imagined: Reflections on Science.* Washington, DC: Smithsonian Institution Press, 1993

Newton, David E. *Chemical Elements.* Venture Books. Danbury, CT: Franklin Watts, 1994

Yount, Lisa. *Antoine Lavoisier: Founder of Modern Chemistry.* "Great Minds of Science" series. Springfield, NJ: Enslow Publishers, 1997

CD-ROM

Discover the Elements: The Interactive Periodic Table of the Chemical Elements, Paradigm Interactive, Greensboro, NC, 1995

INTERNET SITES

Note that useful sites on the Internet can change and even disappear. If the following site addresses do not work, use a search engine that you find useful, such as:
Yahoo:

http://www.yahoo.com

or AltaVista:

http://altavista.digital.com

A very thorough listing of the major characteristics, uses, and compounds of all the chemical elements can be found at a site called WebElements:

http://www.shef.ac.uk/~chem/we b-elements/

A Canadian site on the Nature of the Environment includes a large section on the elements in the various Earth systems:

http://www.cent.org/geo12/geo12/htm

Colored photos of various molecules, cells, and biological systems can be viewed at:

http://www.clarityconnect.com/webpages/-cramer/PictureIt/welcome.htm

Many subjects are covered on WWW Virtual Library. It also includes a useful collection of links to other sites:

http://www.earthsystems.org/Environment/shtml

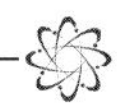

Index

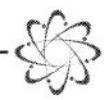